LIFE ON THE INTERNET
GEOSCIENCES
A STUDENT'S GUIDE

ANDREW T. STULL

DUANE GRIFFIN

PRENTICE HALL Upper Saddle River, NJ 07458

Editorial Director: ***Tim Bozik***
Editor in Chief—Science: ***Paul F. Corey***
Editor in Chief—Development: ***Raymond Mullaney***
Acquisitions Editor: ***Dan Kaveney***
Director of Production/Manufacturing: ***David W. Riccardi***
Special Projects Manager: ***Barbara A. Murray***
Manufacturing Manager: ***Trudy Pisciotti***
Buyer: ***Ben Smith***
Formatting Manager: ***John J. Jordan***
Formatter: ***Cindy Dunn***
Proofreader: ***James Buckley***
Art Director: ***Heather Scott***
Cover Design & Illustration: ***Paul Gourhan***

Printed in the United States of America
10 9 8 7 6 5 4 3 2 1

ISBN 0-13-266578-6

Prentice-Hall International (UK) Limited, *London*
Prentice-Hall of Australia Pty. Limited, *Sydney*
Prentice-Hall Canada, Inc., *Toronto*
Prentice-Hall Hispanoamericana, S.A., *Mexico*
Prentice-Hall of India Private Limited, *New Delhi*
Prentice-Hall of Japan, Inc., *Tokyo*
Simon & Schuster Asia Pte. Ltd., *Singapore*
Editora Prentice-Hall do Brasil, Ltda., *Rio de Janeiro*

Preface

Change! As I write this manual the time is 7:22 p.m., April 29, 1996. It might be a little odd that I should bother with such a seemingly insignificant statement, but this brings me back to my first statement, "Change!" We live in a world that is in a constant flux and has always been in such a state. It is a world louder, faster, and more complex than the one other generations have experienced. If you agree with me, you will recognize where we, as a techno-generation, are going and where our parents, as a paper-generation, are coming from. On the evening of this night, April 29, the Internet (a little more than 25 years old) and its most recent incarnation, the *World Wide Web* (less than five years old), are changing at chaotic rates - revealing glimpses of our future. The information and resources that will be available to us tomorrow, April 30th, will have been amplified by this change and the world will be different.

I am writing this manual not to teach you all there is to know about the *World Wide Web* (*WWW* or simply called the *Web*), but rather to help you teach yourself. If I were to give you information without teaching you how to find it yourself, then you would be dependent on me and the bearer of old knowledge. If I and others are successful, your skills in 'surfing' the Internet will allow you to deal with the perpetual change. Eventually, we will laugh at ourselves as we watch later generations skillfully and artfully navigate this sea of information. If you don't believe me, tell a five year-old that you need help operating the television remote. I suspect that he or she would think you silly.

By the end of the manual, you should be comfortable and resourceful in navigating the complexity of the Internet, from its back eddies to its thriving thoroughfares.

There are two main chapters to this manual. In the Introduction, *O' Brave New World*, I will describe the origin and innovations behind the Internet. In Chapter One, *The Mechanics*, I will give you a run-through on the use of a *Web* browser and introduce you to some tools for navigating the endless alcoves of the Internet. The boxed-in and end-of-chapter exercises will give you practice using your browser, reinforce your navigational skills, and give you practice at searching for some of the great resources available on the Internet.

In Chapter Two, *Doing Your Own Tune-Up*, I will explain some of the fine tuning and customizing you can perform on your *Web* browser to make it more responsive to your needs. Additionally, I'll offer some of the essentials to move you from an observer to an *enthusiastic* participant. The boxed-in and end-of-chapter exercises will help you reach out and make contact with others on the Internet.

Writing a manual for geography and geology students is a tough bill to fill, given the incredible breadth that these two disciplines cover. Between the two of them, everything in the world is (literally) a potential subject of interest. I've tried to use examples of *Web* resources that will be broadly interesting, admittedly tilted towards physical geography and geology, a reflection of my specialization. If your interests are primarily in human geography or cartography, don't despair: I've included a number of resources for you as well. Besides, if you'll bear with me, and if I teach you well, you'll be able to find something on just about any topic you can imagine!

Acknowledgments

I'd like to thank my wife, Elizabeth, for the encouragement, the faith, and the unbelievable time she has spent editing this work. She has probably spent as many hours reviewing and providing invaluable suggestions for it as I have spent writing it. I must also thank my partner and mentor Steven Brunasso for his patience, friendship, and unbelievable clarity of vision for the Internet. I am grateful to the following people for providing many helpful comments and suggestions:

Susan Brawley, *University of Maine*
Ronald Edwards, *University of Florida*
Patrick A. Thorpe, *Grand Valley State University*
Robin Tyser, *University of Wisconsin - LaCrosse*
Roberta Williams, *University of Nevada, Las Vegas*

I am also grateful to Ray Mullaney, Editor in Chief of Engineering, Science, and Mathematics Development, who meticulously edited the final draft, and to Kelly McDonald, Executive Marketing Manager, for her insightful comments and guidance. Finally, I would like to thank Sheri Snavely, my editor at Prentice Hall, for listening to my rantings about the direction of education on the Internet, for the opportunity to write this manual, and for her support.

Andrew Stull

I'd like to thank Andrew Stull for all of his hard work on *Life on the Internet: Biology—A Student's Guide*, which provided most of the original text for this manual. I'd also like to thank Dan Kaveney and Ray Henderson at Prentice Hall and Jim Burt at the University of Wisconsin Geography Department for their help and suggestions. Most of all I, too, would like to thank my wonderful wife, simply for being my wonderful wife.

Duane Griffin

Contents

Introduction:

O' Brave New World

There are important things that history can teach you. If you don't learn them you may find yourself walking into problems or even trouble that otherwise could have been avoided. Don't get me wrong, there is a lot of junk out there that is called history, but take a look at it first before you decide you don't really need it.

This thing that we now call the Internet has been around for just over twenty-five years and has been growing and changing ever since it first started. Its age is irrelevant for our discussion here, but the reasons for its creation and growth are quite interesting and helpful in understanding its nature, its terminology, and the culture of the people that have adopted it as home.

The Internet has been likened to something alive because of the way it grows and changes. The best description of the Internet is that it is not a network but a network of networks. In the late 50's and early 60's, scientists and engineers realized the importance of sharing and communicating through their computers. As things go, many different groups attempted to develop a network language and most were successful. The ironic thing was that now there were many different computer networks that still couldn't speak the same language. It was the Tower of Babel situation all over again.

The Internet was born as the solution to this problem. The US government paid for the development of a common network language (called a protocol) which was eventually shared freely. With time, the many isolated networks from all over the world adopted this language (called TCP/IP). There is no central Internet agency and therefore the Internet is independent of governments and regulation. It grows and changes from the common need and commitment of the people that use it.

Admittedly this type of network system isn't the most graceful - but it works. If you saw a diagram of this great big computer network you might easily recognize it as a spider's web. On this web, information can easily travel between any two points along many different paths. This web is called the "Internet," and it was born in 1969. Although the Internet is used very extensively for scientific purposes, as you can tell by the recent explosion of movies and commercials, it is growing for many other reasons.

The Internet is going through many changes in the way information is provided. Although its initial purpose was to allow a distribution system for scientific exchange and research, it gradually took on the role of digital post office (electronic mail and file transfer). As technology changed, the transfer speed and the way we viewed the information changed. Earlier work on the Internet relied on a "command-line interface," where a user simply typed commands from a keyboard and waited for the response. DOS-controlled computers are an example of this. Recent advancements have brought us the graphical user interfaces (GUI). Macintosh and Microsoft Windows are classic examples of systems that use a GUI. These GUI's rely on icons and images to symbolize activities and commands. Actions are initiated with a mouse click instead of a typed command. One of the first commonly used GUI's for the Internet was *Gopher*, developed at the University of Minnesota in 1991. The title *Gopher* comes from the fact that the University of Minnesota's mascot is the *Gopher*.

Gopher is a visual tool for the Internet, where information is presented as a hierarchical listing of directories and files. The users jump from general directories at higher levels to more specific directories at lower levels until they find what they want. Beyond searching for information on a single computer, *Gopher* allows the users to quickly and transparently jump from one computer to another. This method of interaction no longer requires the users to log-on to each computer independently but rather allows them to wander or browse (the *Gopher* people call this process "burrowing") at will among the vast stores of information on the world's computers. Once the information is found, it can easily be downloaded to the user's computer. By simplifying the access and use of the Internet, the number of users and businesses has increased at astronomical rates. Amazing if you think about it! You could be on your computer in Toledo, Ohio, jumping from one screen to another, and without realizing it you might find yourself in Hamburg, Germany, Mexico City, or Tokyo, Japan. Wham! And no airline tickets!

The next major tool for the Internet followed in 1992 with the creation of the *World Wide Web* information format, conceived and initiated at CERN, in Switzerland. Although *Gopher*, as an initial attempt at navigating the Internet, was a good GUI, it lacked visual appeal.

Information on the *Web* is posted as a "page," and it may contain text, images, sounds, and movies. The organization of a page is much like any printed page in a book. There are the visual improvements, but the major innovation is the use of *hypertext*. Hypertext, more specifically hypermedia, is the use of words *and* images as links or connecting points to different text, images, sounds, or movies on other computers throughout the world.

After the introduction of the *Web*, there was a dilemma: a great place to go but no easy way to get there. Kind of like the moon in the 60's. The one thing still lacking was a convenient program that would allow users to access the *Web* easily. This program, *Mosaic*, appeared in 1993 from the National Center for Supercomputing Applications (NCSA). Not only did it allow for the browsing of *Web* pages, but it added the ability to use other Internet resources such as E-mail, *Gopher*, and FTP. With the release of this browser, the *Web* has been growing faster than the speed of light. In 1991, Internet users were around 700,000. After *Mosaic* in 1993, the number increased to around 1.7 million. In July of 1994, the estimate was more than 3.2 million, and by the end of 1995 it was 10 million. This phenomenal growth is attributed to the release of *Mosaic* and sister browsers like *Netscape*. Today the Internet is changing at staggering rates and becoming more readily available to the average person. Just listen to the computer jargon from the magazines, radio, television, or movies.

Chapter 1:
The Mechanics

READY. SET. GO! - Do you remember hearing those words when you were a kid? I used to run foot races with my friends when I was young. The kid that yelled for the start would always jump the line before he called GO. I'd be slow off of the line in some races, fast in others, and occasionally I'd even win. But at the end of the race, it didn't matter because we'd all be happy and we'd do it again - possibly to a different finish.

In the real world, not everyone starts at the same time, but that does NOT determine the finish. In a similar way, many of you reading this manual have a lot of experience with computers, while others have none at all. To begin this chapter you should have your computer set up and operational, including the necessary software and an Internet connection. For those of you who are new to this race, fear not. I have included several resources in the *appendices* at the back of this guide to help you get started. These resources include a description of the necessary equipment (*Appendix I*), a list of the leading software (*Appendix II*), and a list of companies that would be glad to help you connect to the Internet for a small fee (*Appendix III*). After you read the appendices, you'll have an equal chance at the starting line.

There are a staggering number of computers, software, and connections that you can use to get onto the Internet. This manual includes general features common to some different computers and software packages that you may be using. In certain situations, I made a personal choice so that this manual didn't become too generic or void of personality. Most notably, I have illustrated this manual using examples from the *Netscape* 2.0 software on a Macintosh. *Netscape* is a very good browser and <u>free</u> to students, educators, and schools. The basic architecture and features for most other browsers are pretty much the same. Get to know the online manual for the particular browser you're using. It will help with details that differ from the examples I give, and provide a wealth of other useful information as well.

Section 1.1
What your mechanic never told you.

The software that you'll use to access the Net is commonly called a *client* or a *Web browser*. It functions on an information exchange model called a *Client-Server model* (Figure 1). In this way, a client (your *Web* browser software) will communicate with a *server* (other *Web* server software) on the Internet to exchange information. When referring to the *Web*, the information that your browser receives from the server is called a *page*.

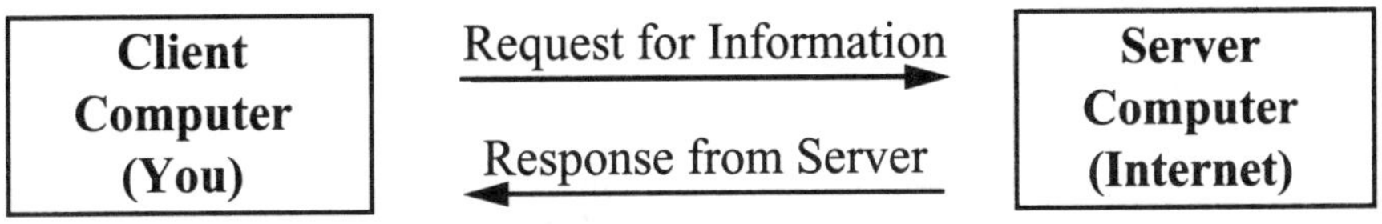

Figure 1. The Client requests information from the server. The requested information is displayed on the client computer.

So, what really appears on these *Web* pages? The best way is to see for yourself. If you can, sit down in front of a computer, start your browser software, and connect to the Internet. Your browser is probably already set to start at a specific page. This start page is often referred to as a *homepage. Web* pages usually include both text and images. Some will also use sounds and videos. The use of different information types is called multimedia. When you choose a page it will be sent to your computer. After the requested information has been sent by the server, your computer will display it for you. As a basic rule, anything that can be saved or recorded onto a computer can be provided and distributed on the Internet through a *Web* page.

GoTo

Because sometimes it is better to begin by running before you know how to walk, I've put this section first. The Internet is large and disorganized, and therefore many people have a difficult time finding their way around it. Finding your way through the Internet and the *Web* is really no harder then finding your way to a friend's house. Suppose that your friends are having a party at their new house and supplying all the entertainment and food. You're familiar with their town so you need only their address to get to the party. Navigating the Internet with a *Web* browser is even easier. If you want to go someplace on the *Web*, all you do is type the address into the browser and connect to the desired page. No map required.

Information on the Internet has an address, just as your friends do. Most browsers allow you the option of typing in an address and going directly to that document. With *Netscape,* there are two easy ways to do this. The easiest way is to use the Open toolbar button. A small window will appear and offer you the opportunity to type in the address of the desired document. A second method, and also one employed by other browsers, is to modify the browser so that Internet addresses are viewable.

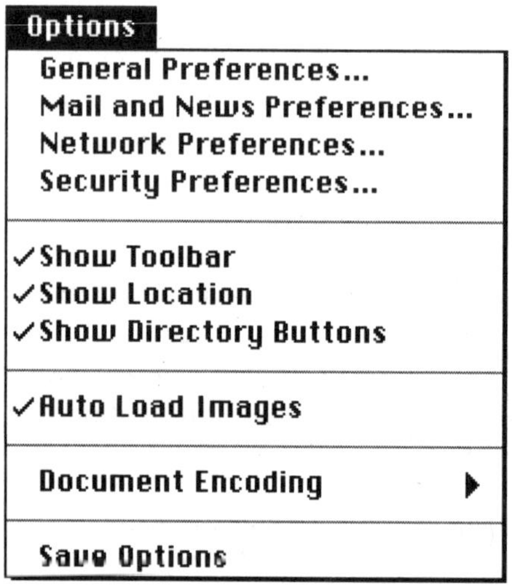

Figure 2. Within *Netscape*, some of the selections under the *Option* menu act as toggle buttons. They can be alternately turned on (with a check mark) or off (without a check mark) by selecting them. Other options bring up submenus to allow further customization.

With *Netscape,* you can open this window by selecting *Show Location* from the *Options* menu (Figure 2). The address window (see Figure 3) will appear immediately below the toolbar buttons. Just type in the requested addresses and press the return button on your keyboard.

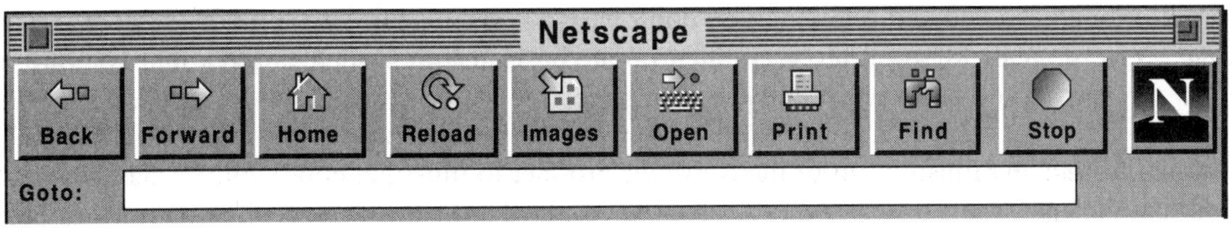

Figure 3. With the *Show Location* option selected in the *Option* menu, the *Goto* window will appear on your browser just below the toolbar.

Now that you know how to do it, here is an address to help you get started and learn more about the *Web*, the National Center for Supercomputing Applications at the University of Illinois, Urban-Champaign. (Appendix II lists some other useful sites.)

http://www.ncsa.uiuc.edu/General/Internet/WWW/Intro/

Navigation

You don't always have to know the address of a page to view it. The wonderful thing about the Web is that you can access (navigate) pages through the use of *hyperlinks*. Hyperlinks are a functional part of moving around the *Web*. You will notice these hyperlinks as colored words on the page. Images may also be hyperlinks. Your mouse is used to select or click on the desired hyperlink. Some *Web* authors write their pages so that their hyperlinks are hidden from you. If you are new, just click on everything. You can't break it.

With recent innovations in writing *Web* pages, other interactions are possible. Clickable images or maps are becoming a prominent way of interacting with *Web* pages. With a clickable image, you have the option to select/click a section on the image, which functions as a hyperlink. This is different from the normal image link because the author of a *Web* page can connect one image to many different places. An example might be a map of the world. The creator of such a document might place a link under each country on the map. When you point and click on a particular country, your browser immediately takes you to different pictures and information about the country.

Here are two addresses that use clickable maps. They are also great places to find science information.

http://ucmp1.berkeley.edu/subway.html
http://www-c8.lanl.gov/infosys/html/periodic/periodic-main.html

Not all browsers support clickable maps but most good *Web* authors will still provide the same choices using conventional hyperlinks.

Hyperlinks are great for navigating the Internet. However, there are many documents that don't incorporate hyperlinks. If you end up on one of these pages, you'll need another method of navigation or you'll be in trouble. This is very similar to driving though a neighborhood, turning the corner onto a dead end and realizing you don't have a reverse gear. The solution, of course, is to *have* a reverse gear. All good browsers incorporate basic features, such as Back and Forward, to allow the user to hop between recently viewed pages.

Although I'm using illustrations of the *Netscape* browser to illustrate points for the manual, you'll find that most browsers differ in appearance only to the degree that their toolbar buttons for basic navigation are in different order (something you can handle). Figure 4 illustrates the basic orientation of the navigation buttons for *Netscape* 2.0.

Figure 4. The *Netscape* Navigator 2.0 *Web* browser offers a wide selection of basic navigation and control buttons at the toolbar. Although this is a representation of the control strip from the Macintosh version, the Windows version is identical in function and appearance. The toolbars from other browsers offers similar functions.

One of the most useful features that you'll use is the Home button. This feature allows you to jump back to your homepage. We'll discuss how to choose and set your homepage later in the manual. Two additional buttons common to most browsers are the Reload and Stop buttons. *Netscape* also includes many other features within its toolbar to give you flexibility in navigating the Internet, but they are self-explanatory and may differ from some of the options provided by other browser programs.

> Don't wait until the last minute. Use one of the addresses I gave you and experiment with the buttons on your browsers. If you want something different, then give the following address a try.
>
> **http://www.yahoo.com/index.html**
>
> *Yahoo!* is a directory of information that was originated at Stanford University. Remember you won't break anything and if you get lost, you can always shut down the program, have lunch, and try again later!

Up to now, everything on a *Web* page has been described as a point, click, and jump. A new feature called *Forms* has added an entirely new dimension to both the user and the creator of *Web* pages. On some *Web* pages, you will find a section where you may be asked to give information by typing from the keyboard. This information might be in the form of a user survey, an answer to a question, or a request by you to a *Web* server. One of the most unique uses of forms that I've seen is on a page for "Mr. Edible Starchy Tuber Head." By selecting from different features on this *form* page, you have the ability to assemble Mr. Edible Starchy Tuber Head in many different configurations. Small things amuse small minds - but when you stare at a computer all day, you need something for inspiration.

Experiment with Mr. Edible Starchy Tuber Head and relax. If you got this far, then you are well off the start line. Not all browsers support forms, so you will need to determine if yours does.

http://winnie.acsu.buffalo.edu/cgi-bin/potatoe-cgiß

Section 1.2

Return to sender - address unknown.

The addresses that you've been using on the *Web* are called Universal Resource Locators, or URL's for short. Each URL has a couple of basic parts just like a residential address. Look at some of the URL's (now you know what they are called) that you've used already. Do you notice any common features or similarities between them?

Here is yet another example of a typical URL and a pretty cool site in itself. There are three basic components to a URL. Compare this with the others I've given to you.

http://kuhttp.cc.ukans.edu/cwis/organizations/kucia/uroulette/uroulette.html

Here are the three components:

protocol	http://
server	kuhttp.cc.ukans.edu
path	/cwis/organizations/kucia/uroulette/uroulette.html

Be careful. Even one incorrect letter will prevent the browser from finding the correct site.

It may seem confusing at first, but think of it as nothing more than a postal address squished together without any spaces. The protocol of the URL denotes how the information is stored. In the exercise above, "HTTP" refers to the protocol or language for which all *Web* servers speak. The colon and slashes are used to separate it from the name of the server. They are not always present with every type of URL. The path describes the location of the *Web* page on the server.

Up until this point, we have been discussing browsers only as an interpreter for documents that use hyperlinks. However, browsers also have the power to link to other, much older, formats of information. Try the following URL's to notice how the information differs.

ftp://oak.oakland.edu/pub3
gopher://gopher.micro.umn.edu:70/1

The HTTP, FTP, and Gopher are not the only resources but they are the three you're most likely to encounter. Have you ever encountered the following Internet protocols?

telnet://
wais://
news:

The last part of the server name defines the domain to which the server belongs. A domain is just a fancy name for a functional network group. The most common domain that you'll see is the .edu domain that encompasses educational institutions.

> Some of the other domains you are likely to encounter are *com*, *gov*, *mil*, and *org*. Here are a couple of examples that you might be interested in. Com stands for "commercial." Can you determine what the other domains represent?
>
> **http://aol.com**
> **http://www.nps.gov/**
> **http://navysgml.dt.navy.mil/patteam/wwwintro.html**
> **http://envirolink.org/**

In your travels, you will eventually jump to a server that is outside your country. Much, but not all, of the information on these servers is in English (so unless you are paying attention you might never know your foreign whereabouts). URL's of foreign servers have an additional section at the end of the servers name. It is a two letter code that denotes the country. Here are just a few examples, but I trust that you can find more.

- .au Australia
- .ca Canada
- .ch Switzerland
- .nl Netherlands
- .pe Peru
- .uk United Kingdom

> I'm sure that most of you have done this but just to catch the loose ends I'd like the rest of you to visit the following site. It was the one we used to illustrate the anatomy of a URL. This *Web* page is a hyperlinked roulette wheel. It is designed to randomly drop you on a URL somewhere in the world.
>
> **http://kuhttp.cc.ukans.edu/cwis/organizations/kucia/uroulette/uroulette.html**
>
> While you are out in the world, see if you can find URL's from other countries. Remember that you can always use your Back or Home button to return to the safety of home.

Bookmarks

Now that you've found so many really interesting pages, how do you find them again a day or a week from now? This is where bookmarks come in. You place a bookmark on the page to which you wish to return. Just to get your collection started, try this address: **http://www.hotwired.com**. It's the on-line version of Wired Magazine and a pretty interesting place. If you like their page, you can save a bookmark by selecting *Add Bookmark* from the *Bookmarks* menu on your browser (Figure 5). Some browsers may be different. *Mosaic* uses the term *Hotlist* instead of *Bookmark*, but the same general idea applies. Now, if you use your mouse to go back to the *Bookmarks* menu, you'll see that *Hotwired* is just a jump away. You don't have to memorize the URL and you don't have to haphazardly jump around until you find it again.

Figure 5. After you've added bookmarks to your browser, they will appear in a list under the *Bookmarks* menu on the menubar. Once you've selected a list of bookmarks, you may use the *Bookmarks…* submenu under the *Windows* menu to arrange and organize them into categories.

Depending on the vintage of your browser, you may have the option of organizing your bookmarks into categories. Additionally, it is possible to export your bookmarks as a group so that you can give them to others. It's similar to trading baseball cards. I won't go into the procedures here, but as with life, it's yours, use it wisely, learn, and have fun.

Travel History

If you've been working along with your manual and your browser at the same time, you've probably been to many places on the *Web*. Obviously, if you want to move back and forth between your selections, you are probably using the Forward and Back buttons on the browsers toolbar. But what do you do if you want to go back to someplace you've visited fifteen jumps ago? Do you press the Back button fifteen times? Well, if you choose to do it that way, *fine*, but it is a little slow. There is a great feature that allows you to jump rapidly to any of the places you've visited along your path of travel. If you are using the *Netscape* browser then click on the *GO* menu option and you'll see a list of all of the places you've visited on your journey. Other browsers have similar features. The simple point is that without using the Back button fifteen times, you can select from a list of recently viewed pages.

Now that you've got the hang of it, give the following URL's a try.

http://espnet.sportszone.com/
ESPN Sports

http://www.foxnetwork.com/home.html
FOX Broadcasting Company

http://www.iuma.com/iuma/indes_graphic.html
Independent Underground Music Archive

http://www.mtv.com/index.html
MTV (Music Television)

http://www.nbc.com/index.html
NBC (National Broadcasting Company)

http://www.nflhome.com/index.html
NFL (National Football League)

http://www.pathfinder.com/
Time Warner Communications (Life, People, Sports Illustrated and more)

When you find something that you think is fun and interesting, save it or give it to a friend.

Activity

365 Adventures Without Ever Leaving Home.

I hope that you haven't waited until this point to travel the *Web*, but if you have, this exercise will get you up to speed. You'll have plenty of opportunity to experience some of the great places that are at your disposal on the Internet. The only thing that you'll need is the knowledge of reading URL's and saving bookmarks.

What you will be doing is riding a roulette wheel to scientific resources on the Internet. This allows you to jump to a random document potentially anywhere in the world. With each *Web* document, check the URL. If you really like the unpredictable nature of randomly jumping around the Internet you might want to also visit these other random jump *Web* pages.

http://kuhttp.cc.ukans.edu/cwis/organizations/kucia/uroulette/uroulette.html
University of Kansas Uroulette page

http://www.netcreations.com/coolroulette/
Cool Roulette

This exercise is really a free ride. You will experience many different places and discover some that have value to you. If you find something that you like, set a bookmark or record the URL so that you can return.

1. Compile a list of the different countries you visit. You have been given a basic set of country domain codes. How many more can you find?

2. How many foreign languages can you find? English is probably the most common but definitely not the only one you'll encounter.

Netscape and many other browsers allow you to organize bookmarks into categories. It is easy to find the information in the user manual. Try categories like: *Fun and Games*, *Magazines*, *News and Weather*, *Outdoors*, and *Science*. Under each of these major categories you can create sub-categories. Using Science as an example, I have sub-categories like: *Climatology, Geography, Geology,* etc...

3. Organize your bookmarks into categories so that they are easy to use. Use the categories that I've described above or make your own.

Section 1.3

Let your fingers do the walking.

OK, now you've learned what the Internet is, how to use a browser, and what resources are available through the *Web*. If you've done your homework, you also realize that there's no way of memorizing the content or location of everything at your disposal. In this section, you'll move from browsing, which is really just the haphazard wandering from one place to another, to active searching for specific information.

The *Web* is an information jungle, and just like any other jungle, it's full of all sorts of valuable stuff. At the same time, it's also full of stuff that you really don't need, so it's also a lot like the bottom of your closet (at least if your closet is like mine), only a whole lot bigger. And like both a jungle and the bottom of your closet, finding your way around without getting too lost can be a challenge.

Sure, you can keep a list of bookmarks, or even create your own portable *Web* page with links to your favorite resources (that's Chapter Two). But eventually you'll find that *Web* resources can be quite ephemeral. URLs change and sites go dead, sometimes alarmingly fast. You may have already encountered the dreaded *Error 404* message, telling you that your browser couldn't locate the server you're trying to reach. Alternatively, you may know exactly what you want, but have no idea where it is. What to do?

Fortunately there are tools to help you navigate the *Web*. There are two types of tools that you'll use most often to search for information. These are *directories* and *search engines.*

Directories

Yahoo! is only one of many directories that are available in the Internet, but it is the best for general topics. As you become more experienced and traveled in the *Web,* you will discover more. I didn't name it and I don't know why it's called that but it is easy to remember. I gave you its URL previously, so I hope that you've saved it as a bookmark. Just in case, here it is again (http://www.yahoo.com/). As I describe this tool, I'd like you to follow along and experiment with it.

Yahoo! is a listing of information by category – kind of like a card catalog. At the top of the listing, there are several very general categories, but as you move deeper into the directory you'll notice that the categories become more specific. To find information, you simply choose the most appropriate category at the top level and continue through each successive level until you find what you're looking for or until you realize you're in the wrong place. Don't worry, I get lost most of the time.

As an example, suppose you were studying a chapter discussing the impact of humans on coral reefs and you wanted to learn more. Within *Yahoo!*, notice that one of the top level categories is Science. Science would seem to be the most appropriate place to find the information so give it a try. After accessing the Science category, you'll notice that it gives a list of many different types of science. So, what is your next choice? My choice is Oceanography but Ecology is also a possibility. Because *Yahoo!* cross references among the categories, you'll find that several related categories will lead you to your desired page.

Much of your success in finding information with this type of tool really centers around your preparation for the search. Often, it is possible to find information on a topic in a category that may at first seem unrelated to your topic of interest. Again, let's take the example of coral reefs and human impacts. Although you may consider this to be a topic to be searched out from a category in science, there are other avenues to consider. Are not coral reefs a common destination for vacationers? When you consider this, you'll realize that categories related to vacations and travel are searchable options. Where are the coral reefs in the world? One of the largest is in Australia; so by selecting Australia and its neighbors you might also turn up something related to coral reefs.

Prepare yourself for a search before you jump into one. In the long run it will save you both time and frustration. Don't be afraid to try some strange approaches in your search strategy. A good technique that I use occasionally is to pull out my thesaurus and look up other names for the word. It may be that you can find a more common form of the word. Think of all the associations that your query may have and give them each a try. You never know when it will turn up a gold mine.

Search Engines

Another more direct approach to finding information on the *Web* is to use what is called a Search Engine. Don't be confused by the term. It is nothing more than a program that runs a search for you while you're waiting for the results. There are many search engines on the *Web* and other places like *Gopher*. *Archie*, *Veronica*, and *Jughead* are search engines that are available to search different aspects of the Internet like *Gopher* and FTP. We won't cover them here, but as you build your experience and confidence, you should give them a try. Some of the *Web* search engines are commercial, and they may charge you a fee to run a search. Just as anything that proves itself to be in demand, the *Web* is turning commercial and you will notice many places will begin as a free service and eventually ask you to pay a fee. Don't worry, there are plenty of free search engines and many more popping up all the time. In fact, *Netscape* and other browsers now include a built-in search engine that's free, provided you're willing to look at a few advertisements while your search proceeds.

A search engine that I use frequently is called *WebCrawler* (http://www.Webcrawler.com/). It's very simple to operate, but as with any search tool, it takes practice and patience to master. Take time now to connect to *WebCrawler*, and we'll take it for a test run. When you first see the opening page, you'll notice that it is very simple. Fear not, behind this page is a very good resource. Reading the instructions on the page will tell you most everything you need to know. Enter a word into the white entry box and press the Submit button, *WebCrawler* will refer back to its database of information and return a page of hyperlinked resources containing the word you entered. When the query results come back to you, notice that they are hyperlinks to somewhere out on the *Web*.

Both as an example of how a search engine works and as a comparison with *Yahoo!,* use "Coral Reef" as topic for the search. Notice that you can set the number of responses that the engine will return to you. How do the search engine results compare with *Yahoo!*? Did you notice that some of your results didn't seem to apply to your topic? If you said yes, then you've notice one of the downfalls of search engines. They are very fast and very dumb. They don't think (that is your job). A search using the term "coral reef" is just as likely to turn up a link to Jimmy Buffett's Coral Reefer Band as a link to coral reef research. To perform an effective search you will need to spend time *before* the search preparing a search strategy. When you perform a search using an automated tool like a search engine you can expect many links to be unrelated to your intended topic – but all in all they are still very powerful tools.

Activity

Finding the World on the Web.

Let's put all the experience you've accumulated to the test. I'll give you several words and phrases, and you try to find something related to them using *Yahoo!* and *WebCrawler* or your browser's search engine.

How many links did your search return for each topic? How many of them are actually related to the topic, and how many are false hits?

History of Cartography	Burgess Shale
Glacier National Park	Paleoclimate
Earthquakes	Clean Water Act
Global Change	Mt. Pinatubo

HINT: Try the following search engine (ORES), devoted entirely to the geosciences:

http://www.gisnet.com/gis/ores/

or

http://www.calweb.com/~tcsmith/ores/

One more challenge.

See if you can find data that might answer the following questions:

Are levels of meat consumption (or traffic accidents, voter turnout, belief in UFOs, or anything else you find interesting) equal across the U.S., or do some regions/states have higher rates than others?

There has been a global trend towards warmer temperatures during this century. Are there regions where temperature trends are neutral, or even negative (i.e., it's gotten cooler)?

> Most *Web* resources include summary information on a topic. These can be good sources of information for course projects, term papers, your own amusement, etc. (see Section 2.4). However, there are also a lot of sites with archives of hard data that have been compiled and made available for researchers. Some of these can be searched or even analyzed online; with others you can download the data files and even special software for analyzing them. These are most appropriate as sources for seminar papers or senior theses, but can also be used for your basic term paper or other course projects.
>
> Take my word for it, there's no better way to impress an instructor than turning in a project based on data analysis. It's also very satisfying to actually produce an original piece of research, rather than just summarizing what others have written about a topic. Finally, the ability to find and analyze data is a great skill to be able to offer a potential employer.
>
> Keep in mind that it's incredibly easy to abuse data and statistics and end up with garbage. Don't be afraid to ask for help. Your instructor will probably be delighted by your ambition and be glad to point out the land mines and suggest approaches you haven't thought of yet.

Chapter 2:
Doing Your Own Tune-up

So far, I've introduced you to the depth and breadth of the *Web* and how to successfully navigate it. Hopefully you've had a chance to explore some of the amazing places on the *Web*, found a few favorites, and honed your skills at finding information you need quickly and efficiently. This chapter will cover some advanced techniques to help make your *Web* experience even more enjoyable and profitable.

I've illustrated this manual with *Netscape Navigator 2.0* because, at the time I started writing this, it was the best browser on the market, and—even better—it was free to anybody involved in education. But choosing a browser is sort of like choosing a new pair of shoes: the important thing is finding a pair (or a browser) that fits your preferences, needs, and style, not your roommate's or mine.

There are several browsers listed in Appendix II, most of which are available free to students, at least for a trial run. Try several, and pick the one you like the best. Whichever browser you end up using, you should find the following descriptions and techniques useful. Remember, the *Web* adventure is not in *what* you use, but *how* you use it.

> So that you have something for comparison, here are the homepages for the two most prominent *Web* browsers, *Mosaic* and *Netscape*. They are free if you are a student or work with a school. Within these pages you will also find the manuals for each program.
>
> **http://www.ncsa.uiuc.edu/SDG/Software/Mosaic/NCSAMosaicHome.html**
>
> **http://home.netscape.com/**

Section 2.1

I prefer it *my* way!

Know your options and you'll save both time and money. On the Internet, and as a matter of fact, in life, "Time is money." According to Einstein, "Time is relative;" so I guess you could also say - "Time is relative to the amount of money you have." I wonder if Einstein thought of this? No, I'm not rambling. Here is where I get to the point. If you are like me, you don't have much money, and time is very relative if you have to use a modem and a service provider to surf. If you know what your browser can and cannot do, and if you configure the options for your browser so that it is most efficient for you, you will save oodles of time. With *Netscape*, under the *Options* menu (Figure 7) you are allowed several choices to customize the appearance and function of your browser. I'll say it again, don't wait for me. Try these techniques as you go along and jump around to the information that is important to you.

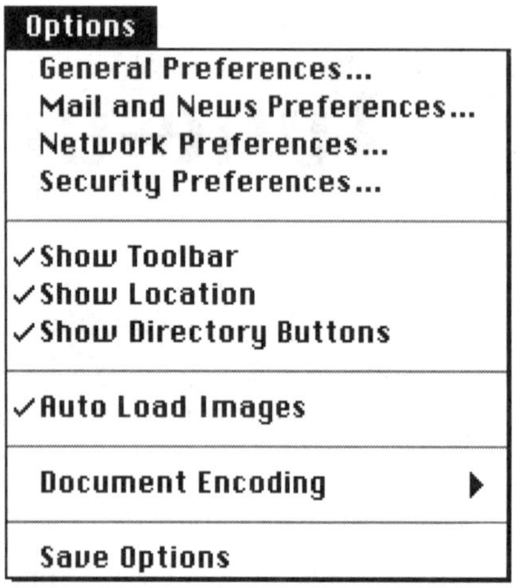

Figure 7. Many of the general options are listed below in the *Options* menu on *Netscape,* but if you wish to do a more detailed configuration, you can find more options under the *Preferences...* submenus.

I described some of these options previously, and some of you may have already experimented with the others, but here is an explanation for the rest of you. Notice that there is a check mark by some of the options. Look at *Auto Load Images* as an example. When the check is present, images will be automatically loaded. If you select this option again, the check mark disappears and, subsequently, pages will load the text without the time-consuming images. Give it a try. If you're trying to connect to the Internet through a modem, as I frequently do, you've probably noticed that some of those images, sounds, and videos take a long time to download to your computer. This, in itself, isn't a big deal until you consider that some pages are crammed with images and downloading them requires a lot of time. This really digs into your wallet if you pay an hourly connection rate to your service provider. What I do to save some money and frustration is to turn off the automatic image load on my browser until I get to a place of interest. In a similar fashion, the other options from this pull-down menu work much the same way.

If this isn't enough control for you, there are other changes you can make. Notice the *Preferences...* submenus under *Options* (Figure 7). When you select one of these, you will be presented with a series of dialog windows, where you can make even more customizing options. If you don't like the way the font, color, or style of the text appears on your browser, you can change them within this dialog window. Browsers typically provide you with many features to allow you to change the visual appearance of the information on the viewed page.

After opening the *General Preferences...* dialog window, a window containing several "folders" of options will appear for you to select from. Each of these options will take you to a different preference window. Figure 8 shows the folders you have to select from when customizing your browser. This dialog window usually opens to the category you were last modifying, so don't be disoriented when it looks different each time

One of the first preferences that you might want to modify is the location of your homepage. A homepage is sort of like home plate in baseball. It is the place where you begin your trip around the bases. If you're working with a browser that hasn't been customized before, it probably has the homepage set to the company that made it. If you're working on one of your school's computers, then the homepage may already be set to the school's page. If neither is the case, then within the *Appearance* preference folder (Figure 9) you have the option to designate a homepage. By now you've probably found something on the *Web* that you are willing to call home. Enter the URL as illustrated below and every time you start your browser or select home from the toolbar you'll end up at this URL. Later in this chapter we'll discuss how you can make and display your own homepage. Stay tuned. Notice that you also have the option of modifying the appearance of your toolbar.

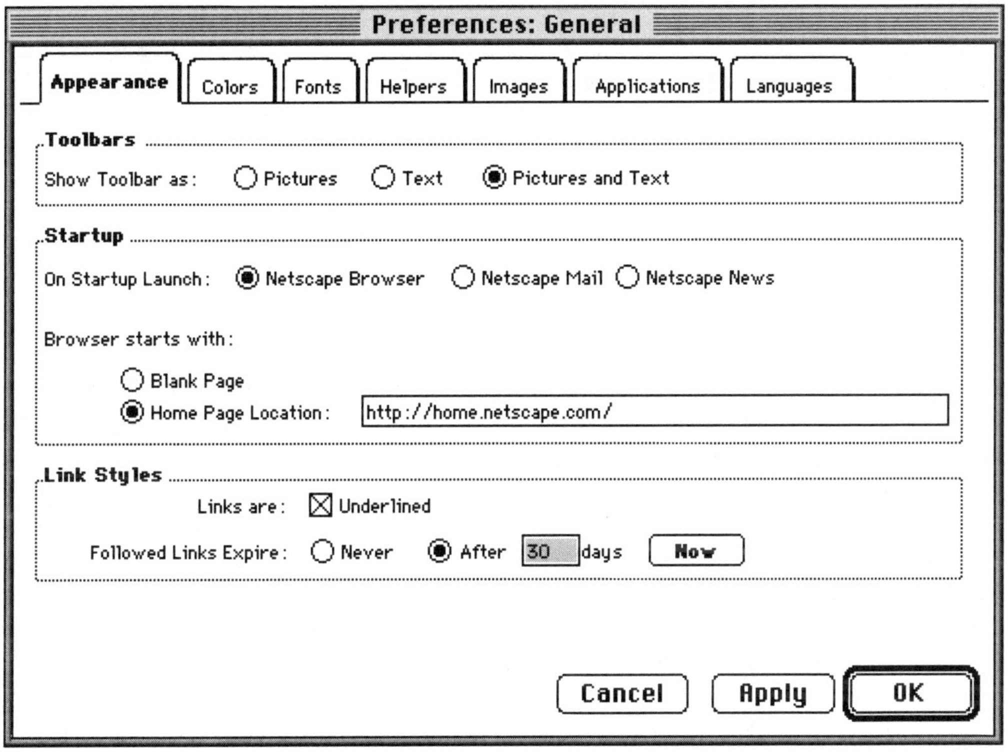

Figure 8. There are many preferences available to you from within the *General Preferences...* menu.

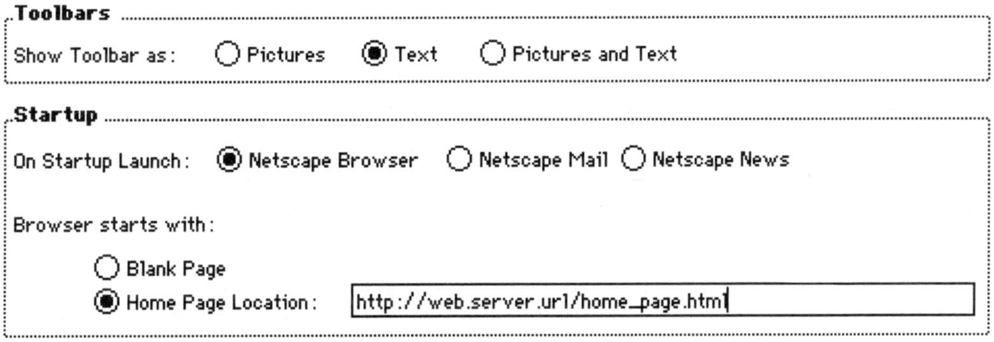

Figure 9. Your browser is easily customized to display your toolbar and homepage preferences.

On the *Appearance* preference folder you also have the option of customizing the appearance of your hyperlinks. Hopefully, you've noticed that hyperlinks are normally blue (this is if you or someone else hasn't changed the default color). Once you've accessed them, they change to purple. This follows the bread crumb principle. It lets you know that you've already been down that link and that therefore you should select a different path. You can set the length of time for a hyperlink before it changes back to the default color. As illustrated in Figure 10, I have my link expiration set to one day. Notice that you can have the browser underline the hyperlinks or leave them plain.

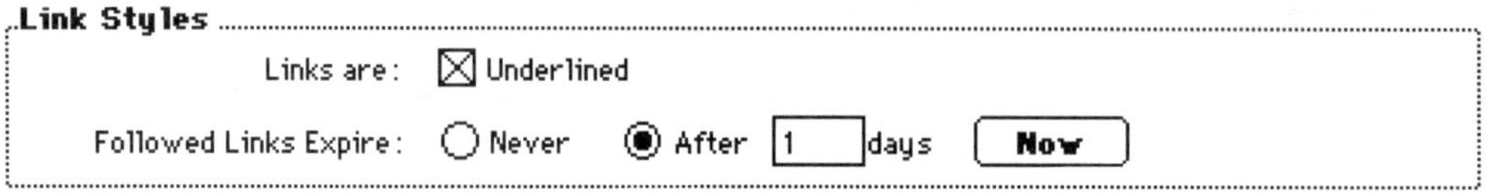

Figure 10. The appearance of hyperlinks should be set to match the degree to which you travel the Internet. If you travel frequently, you may find that your hyperlinks will always disappear and are therefore set too high. If you travel very infrequently, you may find that it is best to never expire a traveled link.

If you select the *Fonts* preference folder you will see a dialog window (Figure 11) that will allow you to modify the font style and size of the text as displayed by your browser. If you typically have a difficult time reading small text from a computer screen, then you may want to increase the size or style of the font.

Figure 11. Selecting a font of appropriate style and size will help you prevent eye strain and muscle fatigue from leaning over your computer to read.

As with fonts, the default text, hyperlink, and background colors can be modified on your browser to meet your individual desires (use the *Colors* folder—Figure 12). My suggestion is for you to set the colors so that you maximize the contrast between the components. Of course, you may choose to color your text the same color as your hyperlinks just to cause you more chaos in navigating a page.

I have not covered all of the options available to you for customizing your browser. An extensive explanation is available to you through the *Netscape* homepage. If you are using another browser, you will find that they have similar options, and on occasion some browsers may give you more power to control your browser's appearance. Again, it doesn't matter which browser *you* choose to use. All I have to say is, *To thine own self be true.* Okay, I want to say one more thing - *May the force be with you, young Skywalker.* I've always wanted to say that.

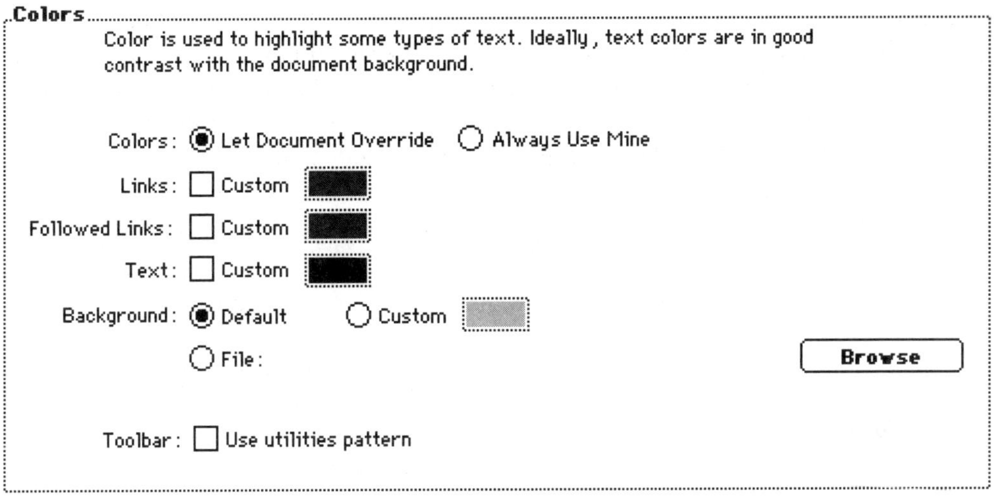

Figure 12. Customizing the colors of the components on your browser will help you quickly identify hyperlinks and navigational paths as you travel.

Section 2.2

Do you get the news delivered?

Contrary to what you might be thinking, learning is a give-and-take activity. It is an interchange of ideas. Effective learning involves communication between all those involved – instructors and students. This means that students need to learn how to communicate among themselves. Within any learning environment, everyone brings a different and unique perspective. With the interchange of ideas, experiences, and understandings shared by each person, the whole community benefits. This brings me to my next point. *Netscape* and many other browsers have incorporated the ability to allow you to communicate through e-mail, newsgroups, and on-line interest groups.

E-mail

As we've discussed earlier, e-mail is the electronic exchange of mail among people. This exchange is from one person to another person. E-mail follows a similar Client-Server model as do *Web* browsers and *Web* servers. Your client software asks a server computer for new items of mail addressed to you. Your mail server is operated and maintained by the administrator of your campus or service provider. Where *Web* servers speak a computer protocol called HyperText Transport Protocol (HTTP), mail servers speak a computer protocol called Simple Mail Transport Protocol (SMTP). Another analogy for you: the server is like your mailbox. The mailbox is where mail addressed to you is stored until you pick it up. You use your mail client to get your mail from your mailbox.

E-mail links are built within many *Web* pages and allow you to send directed correspondence to other participants through the Internet. Although *Netscape* allows you to send e-mail to others, it doesn't allow you to receive e-mail directly. You can configure the browser to receive mail from a server machine that acts as a mailbox for you, and you can configure it through the *Servers* folder of the *Mail and News Preferences…* submenu. However, this will likely involve the cooperation of a friendly campus computer person, so we'll leave the details to them. There are other pieces of software that allow you to receive e-mail. *Eudora* is one of these pieces of software; you may have heard of others. As with *Netscape*, *Eudora* has a free version, and it is available for both Macintosh and Windows. Manuals that describe its use are common and easily available over the Internet. Some commercial browsers come with the ability to both send and receive e-mail but they may cost a little more money.

If you follow this URL, you'll end up on the QualComm homepage where you can download either the free version of *Eudora* or get information on the more powerful commercial version. You can also download a copy of the basic user manual.

http://www.qualcomm.com/quest/QuestMain.html

The following URL's are *Web*-based *Eudora* manuals. Remember to save a bookmark.

http://www.cdtl.umn.edu/eudora_http/contents.html
http://cutl.city.unisa.edu.au/eudora/contents.html

Access to an e-mail server is easily available. If you are at a campus that has provided you with Internet access, you should also have the ability to apply for an e-mail account through your campus computer administration. Again, if you haven't enrolled in an enlightened school, then you have the option of applying for an e-mail account through an independent service provider. In fact, if you've already gone through a service provider for access to the Internet, you probably already have e-mail capabilities. There are two things that you'll need to know to properly configure your browser for sending e-mail: (1) your e-mail address and (2) the name of your mail server. You can enter your e-mail information through the *Mail and News Preferences...* submenu under *Options*. Once you open the preference dialog window, select the *Identity* folder (Figure 13a) to set up your personal e-mail information and the *Servers* folder (Figure 13b) to identify the computer that will act as your mailer. Once you've entered the correct information and restarted the browser program, you should have the ability to send e-mail. Think about it! It means saving postage and long distance phone calls.

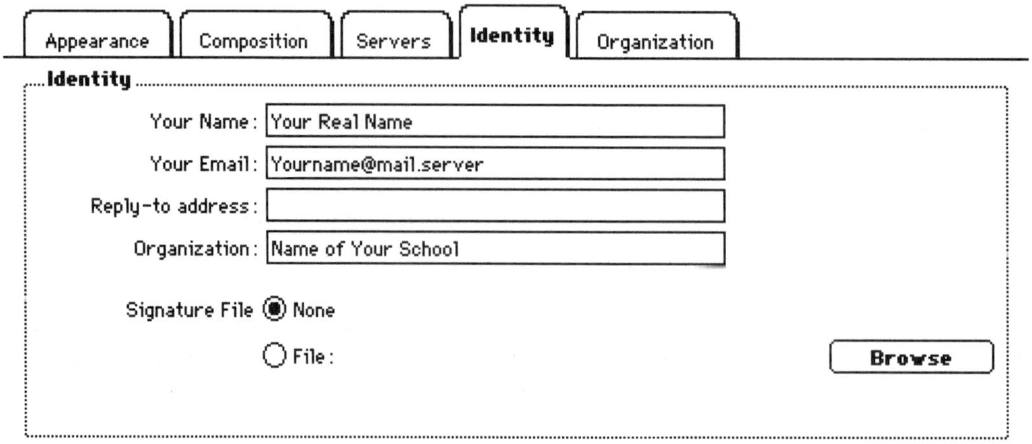

Figure 13a. Insert your personal e-mail information in this folder.

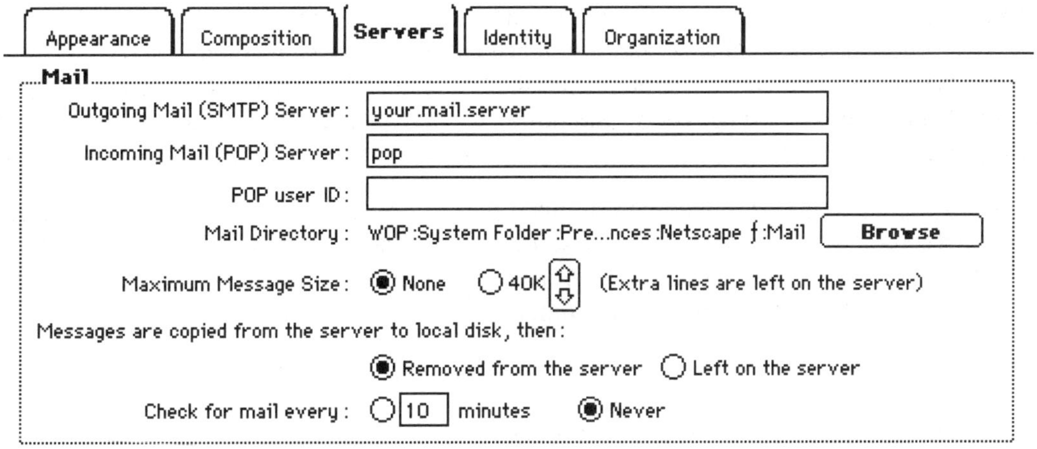

Figure 13b. The information about your particular mail server and e-mail address must be entered before you have the ability to send e-mail to others through *Netscape*.

E-mail addresses typically take a very distinctive format. Here is the format of a typical e-mail address: NAME@HOST.DOMAIN. It is not necessary to have a full name for the NAME part of the address, and in fact, some addresses use only numbers to represent an individual. The @ symbol always follows the individual's name and then the name of the mail server computer (HOST). The domain in the e-mail format, just like the domain of the URL format, is used to denote the affiliation of the user. Notice that there are no spaces anywhere in an e-mail address.

Here is an example of a typical e-mail address. There are three basic components to a typical e-mail address.

D Griffin@geography.wisc.edu

Here are the three components:

individual	DGriffin
mail server	geography.wisc
domain	edu

E-mail addresses are not case sensitive. I use capitals so that addresses are easier to translate.

Newsgroups

A second feature incorporated by many browsers is the ability to exchange ideas over something called the Usenet News, or Newsgroups. Usenet is actually a separate network from the Internet, but accessible from it via most *Web* browsers. A newsgroup is a group of people that connect to and participate in a specialized discussion. These groups are open forums, and all are welcome to contribute in the interchange. Some are moderated by an individual or individuals that will post items for discussion and moderate the electronic brawls that ensue. Others are free-form and function much more like a street fight. The bottom line with all newsgroups is to have a central place where people can bring to the table new ideas and perspectives. No, not all newsgroups are places for disagreement. Many are very good, with courteous people and polite conversation.

Newsgroups are nearly the opposite of e-mail. Where e-mail is sent and posted for a single person to read, news postings are sent to a common place for all to read and comment. By now you've noticed that I like using analogies. Here is yet another analogy that will help you understand the nature of a newsgroup. If e-mail is like the postal service, newsgroups are like coffeehouses on open microphone night.

Okay, so what is the purpose of the newsgroups if people just contribute ideas and comments to a mass discussion on a particular topic? Newsgroups are a place where you can go and ask questions, get ideas, and generally learn about specifics. Select a group that has an interest common to yours or a topic that you are trying to learn something about. You'll find that there are many newsgroups that focus on issues of science (and everything else that a human mind could think of).

The names of newsgroups are usually descriptive of their discussion topics. There are several major newsgroup categories. The one you'll use most is the SCI group. This group contains most of the general science discussions. Other groups include GEOSCI for Geology specific topics, COM for computer topics, and ALT for alternative topics (this is usually where you find most of the really interesting newsgroups – ALT.BIGFOOT.RESEARCH, for example). I've included a list of science and Geography and Geology-related newsgroups in Appendix IV. You'll find some of them interesting and helpful as a supplement to your studies in Geography and Geology.

Here is an example of a typical newsgroup URL. Newsgroup names are divided into several descriptive words separated by a period and organized into a hierarchy. Notice that newsgroup URL's don't use the double slash (//).

news:sci.geo.geology

Here are the three components:

protocol	News:
top category	sci
sub category	geo
sub category	geology

To configure your browser to connect to a newsgroup, you will need the name of the server computer that handles newsgroups. Figure 14 illustrates the preference dialog where you will need to enter your news server name. This may or may not be the same computer that handles your e-mail and *Web* services. News servers use yet another protocol beyond HTTP and SMTP. News servers use a protocol called NetNews Transfer Protocol or NNTP. To keep away feelings of being overwhelmed, on donut day, take a couple of the really sugary donuts down to your local hacker and don't forget the coffee. There is nothing you can't get out of a hacker with a little sugar and caffeine.

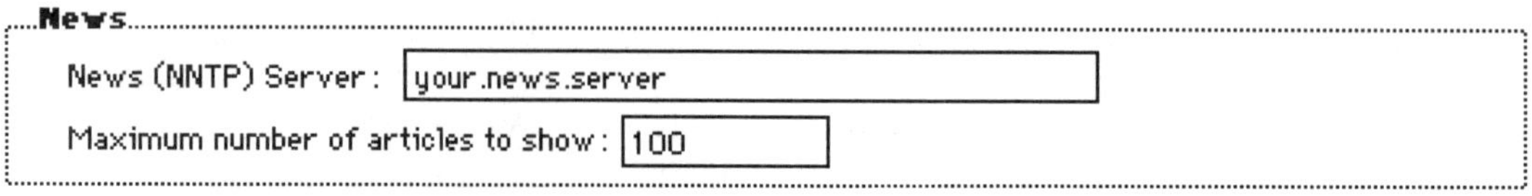

Figure 14. With *Netscape*, the dialog window where you will enter the name of your news server can be found in *Mail and News Preference* submenu under the *Options* menu.

Once you have an e-mail address, the name of your mail server (SMTP server), and the name of your news server, you're set. If you are using *Netscape*, you can configure your browser by selecting *Preferences* from the *Option* menu. Within the preference window that appears, you must select *Mail and News* from the pull-down option box and then enter your information as illustrated in Figures 13 and 14.

The only way to really understand what newsgroups are like is to try them. Appendix IV includes a list of newsgroups you might want to check out. For your first time, you should just read the postings and follow some of the conversations. Connections or threads develop between postings as people add comments to earlier postings. As you become familiar with the topic of discussion, you might post a comment yourself.

To get the real flavor of a newsgroup, you need to spend more than just a couple of minutes viewing it. Choose a newsgroup of interest. Within that newsgroup, select a topic that is being discussed and follow it for a week. You'll notice that some threads generate no replies at all. Others grow and mutate, eventually spawning new threads that mutate again or gradually just fade away. Watch long enough and you'll see at least one example of bad "netiquette." Pay attention to how stupid and/or rude it makes the sender appear. Bear that in mind as you post. After all, you never know who might be watching (see "A Final Word" at the end of this chapter).

Usenet News is not the only way to participate in group discussions via computer. There are also bulletin boards and MOOs, MUDs and chat rooms where you interact directly with other users (assignment: find out what these terms and acronyms mean). A hybrid between e-mail and Usenet is the mailing list. These are discussion groups in which members ("subscribers") post messages that are distributed via e-mail to all of the other members. Subscribing is easy and free (see below).

Some lists (and newsgroups) are very active, posting dozens of messages each day, while others are quiet and may distribute a message only every month or two. Similarly, some lists are moderated (i.e., posted messages are reviewed to ensure that they are appropriate to the list) before being passed on to subscribers. Others are not. A few lists are available both by subscription or through Usenet. An example is the GEOGRAPHy List, a global mailing list maintained in Helsinki.

To subscribe to GEOGRAPH, send this message:

SUBS GEOGRAPH your full name

e.g.: **SUBS GEOGRAPH Jane Dough**
to: **LISTSERV@SEARN.BITNET**

Don't write anything in the subject line, and turn off your signature if it's automatically added to outgoing messages.

Alternatively, you may be able to view the list (depending on your service provider) in Usenet as:

LIST.BITNET.GEOGRAPH

If neither of these works (because, as I keep saying, things change), try using the search tools you've learned to find the GEOGRAPH list.

Section 2.3

There's no place like home.

There *is* no place like home. You may or may not have found some place on the *Web* that you are willing to call home, but if you haven't found one, then hopefully you'll be able to use the following description to create one for yourself. It is not as simple as browsing the *Web* but it's not very hard either. If your campus has a *Web* server, then they may have a place for student homepages. Ask around to see if you are allowed to keep a homepage there. If you can't get access through your campus, you still have other options. Some of the pages that you view on your browser reside on a *Web* server, but you can also create a file on your home computer or even on a floppy disk that will act as a homepage for you. Admittedly, in order for your homepage to be on the Internet, you'd need access to a *Web* server; but it's not necessary for the basics that I'll describe here. Another option is to use a private service provider to gain access to the Internet. Some service providers rent space to keep homepages on their server. One of the latest things to be expanding is the idea of the Internet Coffee House. There are several that have just opened up. They'll serve up great coffee and provide you with access to the *Web*. Some have even gone as far as renting space for homepages. Search one out.

Homepages

Homepage this and homepage that. What good is a homepage? You will find other uses, but here are a couple of reasons I keep a homepage. I use a homepage to organize my important and frequently accessed links and bookmarks to the Internet. Most importantly, these are links and bookmarks to science and geoscience resources all over the world. It's better than a library card. I also keep links to on-line magazines, the weather at my favorite camping spots, movie, music, and book reviews, the current news reports, my school's homepage, my friends' pages (only the cool ones), the manuals for my e-mail and *Web* clients, and my favorite search engines and directories. I've even seen students create hyperlink pages that they gave to their instructors as homework assignments. Imagine turning in an assignment where your instructor could use a *Web* browser to link to your references and resources on the Internet. I wanted to keep a picture of my mom on my homepage but she didn't like the fact that millions of people could see it. You'd better ask yours too. My homepage also has a little about me. I'd like to think that it reflects a personality that's not boring, dull, or generic (although some have said that that *is* my personality).

Now, where is yours? Actually, I've already made one for you. All you have to do is add the stuff about yourself and the links that you think are cool and interesting. Once you've done that, it's all downhill. If you look at Appendix V, you'll see that there is a lot of gibberish that may or may not look like computer programming to you. This is the homepage that I've written for you, and it's written in HTML. HTML is a language that *Web* browsers use to read hypertext documents. It stands for HyperText Markup Language. It's a very simple computer language. In fact, it's so simple that it ranks closer to slang than language. Although this homepage in its present state is quite generic, you'll have plenty of opportunity at the end-of-chapter exercise to liven it up. I don't want you to do anything with it yet. I just want you to see what HTML looks like. We have many basics to discuss before you'll do anything reasonable with it.

> Just to give you some ideas of what homepages are like, you can use the following URLs to see what others have done. Some of these may be appealing to you while others may be excessive. Homepage design is a personal choice.
>
> **The People Page**
> **http://www.nhmccd.cc.tx.us/people/rhese/people/**
>
> **Make A Friend On The Web**
> **http://albrecht.ecn.purdue.edu:80/~jbradfor/homepage/homepage.html**
>
> **A Field Guide To Home Pages**
> **http://nearnet.gnn.com/gnn/netizens/fieldguide.html**

HTML

The gibberish that you saw on the template homepage in the appendix is really pretty easy to understand. There are only a couple of things that you need to remember in order to read HTML. If you ignore the HTML gibberish that you don't quite understand, you'll see that the rest of the document makes sense. You can do this by alternately viewing your homepage file through your browser and then through your word processor. You don't need a fancy program to write HTML, just a word processor and an understanding of the code. Now look at the page with the gibberish. Notice that the HTML gibberish typically, but not always, comes in sets of two. At the very top you'll see a tag like this: <HTML>. A *tag* is an element of HTML code. It is added to the page to emphasize the text that will appear on the viewed *Web* page. Tags always have a angle bracket (<) at the beginning and an opposite angle bracket (>) at the end. Here are a couple of examples: <BODY>, <HEAD>, and <TITLE>. The other part of each tag set has a forward slash (/) preceding the word. Here are the other members of the tag pairs: </BODY>, </HEAD>, and </TITLE>. Tags, when used in sets or pairs, surround a section of text and define the beginning and end of an HTML section. Yes, it is possible to use tags incorrectly, just as it is possible to use English incorrectly. Here is an example of the basic HTML page that any properly coded page would conform to.

```
<HTML>
    <HEAD>
      This is where a description and purpose of the page is written. The material that is here
      is not intended to be viewed on the browser. It is only used to document the intent of
      the page.
    </HEAD>
    <BODY>
      This is where the material that is to be viewed through the browser will be written.
    </BODY>
</HTML>
```

All properly coded HTML documents are coded with this basic format. Notice the beginning and ending member of each tag pair. Although tags usually operate in pairs, some are used as solitary tags. Here are important examples of solitary tags that you should be aware of because they appear in your generic homepage. When you want to separate sections of text with two spaces you might use the paragraph tag, <P>. Two line break tags, <BR>, will also do the same thing. If you wanted the browser to make a horizontal line across the viewed page, then you would use a horizontal line tag, <HR>. If you're paying attention to the tags, you'll notice that the tag names are abbreviations of what they do. Remember, you don't really have to absorb all this language stuff now. I made a link on your homepage that jumps you to several HTML resources.

If you really want to jump into this HTML thing, then here are two more resources that will help you go further. I have tried to find the resources that will be most beneficial to beginners. Some of these URL's may already be included on the homepage I created for you.

Homepage Construction Kit
http://nearnet.gnn.com/gnn/netizens/construction.html

A Field Guide to Home Pages
http://nearnet.gnn.com/gnn/netizens/fieldguide.html

Many other people have already written cool homepages and you could use theirs for a template if you don't like the one I wrote. One of the best things about writing HTML is that you really don't have to write it yourself. *Netscape* and many of the other browsers provide a great tool. If you select the *Source...* sub-menu under the *View* menu at the top of your screen, the actual HTML code of the page you are viewing will be displayed for you. All you have to do is find the cool and interesting things that others have done and copy the code and modify them to meet your personality.

Hyperlinks

One very important element in HTML is the code for hyperlinks (jumping points within *Web* pages). You know most of what you need to know to create hyperlinks. The HTML tag for a hyperlink is called an anchor and it uses the tag pair <A> and </A>. There are two main components to an anchor tag. The first element is the URL and the second is the word. Here is an example of an anchor tag.

```
<A HREF="URL">
    Word
</A>
```

This is a little more than the <A> and </A> that I described earlier but it's still not that hard. I'll dismantle it for you and you can see how it works. The first line of the example above begins the code for an anchor. Notice again the <A>, but also notice that there is a lot more garbage thrown in between the A and the last angle bracket. You should also notice that this garbage is a URL. This URL tells the browser the location of the requested document. The second line is the text as it will appear on the HTML page (the button, so to speak). Because it will be a hypertext link, it will appear as a colored link by your browser. The anchor tag set is completed with the ending anchor tag, </A>. As it operates on an actual *Web* page, if the user selects the hypertext link, then the user will be sent the document specified by the URL.

Now for an exercise. Use your word processor and type in the basic page elements that I gave to you earlier (<HTML> and </HTML>). Within the body of the page I want you to type in an anchor tag set using a real URL. Use one or both of the examples provided below. Once you have this page, save it with the name TEST.HTML and view it with your browser.

A Beginners Guide to HTML
http://www.ncsa.uiuc.edu/General/Internet/WWW/HTMLPrimer.html

How to Publish On The *Web*
http://www.thegiim.org/

See step #2 in the section below for help in finding and displaying this file. Once you've got this page up and running on your browser, try the hyperlink. Does it work? You should notice that these URLs will take you to resources that will tell you more about *Web* publishing.

Home

So, you have a homepage, where do you put it to make it work? Well this depends on whether you will have access to a *Web* server or not. I'll assume that you don't and give you some instructions that will still help you use it to your advantage. Once you've completed your homepage, you'll have a file on either your hard drive or a floppy disk. The advantage of having your homepage on floppy disk is that you'll have a portable homepage that you can use on any computer. I'll give you a five-step procedure for finding and setting your default homepage to the one you've created. (Keep in mind that the details may be a little different for your browser, so check the on-line manual or help.)

1. After you finished the end-of-chapter exercise, you should copy your intended homepage to either a floppy disk or your hard drive.
2. With *Netscape* you should notice that there is a command under the *File* menu called *Open File....* If you initiate this command, you'll get a dialog window where you can designate the file you wish to open. The trick to this is that this file isn't on a server. Select the HTML file from either your floppy drive or hard drive and press the Return key on your keyboard. Because your file is written in HTML your browser will open up and display it for you.
3. If it isn't set already, change your *Options* setting so that you can view the URL's of the pages that display on your browser. When this is set properly, you'll notice that your file's URL is listed something like this: file:/// ..../file.html. Notice that the protocol is not HTTP and that you may have more than two slashes.
4. Record this URL on a piece of paper so that you can refer to it in the next step. I would ask you to remember it and copy it to step 5, but every little slash and colon is important. So, this will save you from a possible SNAFU (Situation Normal. All Fouled Up. I learned this from a friend in the Army).

5. If you remember from earlier in this chapter, I described how you can change the homepage designation for your browser. Go back to this section and use the description to open the *General Preferences* section of the *Options* menu. Open the folder for *Appearance* and enter the URL that you wrote down from step 4 into the homepage entry window.

Now that you've performed these steps, your browser should automatically jump you onto your homepage when you select the Home button from the toolbar. Give it a try. The procedures that you've just performed have allowed your browser to memorize the location of your homepage file. If you move it or erase it, then you'll have to do this all over again. Builds character - right?

Run Wild

There are many more tags than I have the space to describe here or, for that matter, that you probably want to know right now. They are not hard to learn, and in fact, there are places on the *Web* that offer on-line tutorials on HTML design. Once you have your homepage operational, you'll see that I included links to some of these places for you. As you learn more about the Internet, the *Web*, and HTML, you'll discover ways to add pictures to your pages. Later you'll even begin adding sounds and maybe a video clip or two. We are at the beginning of a new way of learning and communicating. The tools available to you at this moment are only a hint of what will be available to you in a year. If I could give you one bit of advice on your journeys through this new world, this *World Wide Web*, it would be to *watch*, *look*, and *listen* to your world.

Activity

Make yourself at home!

The activity of the day is to make your own homepage with bookmarks. One of the greatest problems that I have as a teacher and administrator of a student network is the gradual and insidious buildup of cluttersome bookmarks. The situation is obvious: each student changes the browsers on the computers in my classroom to meet his or her personal needs. I never know what to expect when I need to teach a class or use a browser. Students with much less experience are confused and disoriented. I have, to some extent, come up with a solution.

The gibberish that you will see in Appendix V is my solution. It is a template for a student homepage to be kept on a floppy disk. The added value in this is that you can also use this generic homepage as a portable resource disk on most networked computers. The only drawback is that if you regularly jump between Macintosh and Windows, you'll need two different floppy disks, one in each format.

You should know a couple of conventions so that you will understand what you need to do. I have used a *large italic font* to represent the text that you are to replace with your information. HTML tags are represented in a <SMALLER FONT, ALL IN UPPER-CASE AND WITH ANGLE BRACKETS>. Text in regular size and font does not need to be changed.

To create a homepage, simply use any word processor or editor to type the information in Appendix V (don't worry about matching the fonts) and save it to a floppy disk. Give the file a name with the suffix **.html** (or **.htm** for Windows). If you're using a word processor, be sure to save the file as an ASCII file, rather than in your word processor's format!

Section 2.4

A Final Word (of caution)

I can't in good conscience leave you without mentioning a few words of wisdom and warning concerning the *Web* and your educational career. In short, they boil down to "don't let the former destroy the latter." "What?!?," you say..."The *Web* is the most incredible information resource ever invented, how could I possibly *ruin* my educational career?" Yes, the *Web* is a fantastic resource, and it *should* further your education, not hinder it. Unfortunately there are a lot of ways to use the Internet incorrectly, all of them easy. I've listed three of the most egregious blunders and how to avoid them.

Mistake # 1. I live on the Web. The *Web* is virtually infinite, and constantly expanding. No matter how hard you try, how many hours you spend, you'll never be able to check out all the sites, or even all the sites you might be interested in. If you're susceptible to non-linear thinking, like I am, the *Web* is the ultimate temptation to distraction. You sit down with every intention of finding something you need to finish that paper on earthquake mitigation that's due tomorrow. The next thing you know, the sun's coming up and you've spent the last ten hours learning more than you ever wanted to know about alien abduction, though you did find time to read all the postings to alt.cesium (one of the weirder newsgroups, definitely worth checking out). A closely-allied pitfall is Internet addiction, which is recognized by some psychologists as a real disorder. Don't mistake virtual reality with the reality in which your grade point average resides.

Stay on task, and limit your browsing. If you have to, stick a post-it note on your monitor to remind yourself what you're there for. If you find yourself spending more than a few hours a day on the Net, seek help.

Mistake # 2: If it's easy to get, it's gotta be good. After your first successful hunting foray in cyberspace you may become drunk on the ease with which you found your quarry. A few mouse clicks and keystrokes are much easier than spending hours in the library searching for printed information on a term paper topic. But all information is not created equal. Most books, and all papers and articles in scientific journals have to pass through a process of peer review, in which other professionals review the manuscript, make comments and suggestions, and generally let the author know about mistakes, inaccuracies, or other problems. The author is then obligated to take these into consideration and revise the manuscript before it is printed. This is academic quality control, and it serves a very useful function.

On the other hand, anybody can publish just about anything they want to post on the *Web*. This is one of the great strengths of the Internet, but it can be a problem if somebody has posted an article full of factual errors, sloppy logic, or outright garbage, and these are unfortunately common. There is no "Good *Web*keeping Seal of Approval" to let you know what's good and what's bad. You're left to your own devices, and if you blow it, the grade you get is yours, not some anonymous *Web* denizen's.

Stick to reputable information sources, read critically, and cite your sources, just like you would if the information came from a journal or a book (i.e., provide your reader with the name of the author, the title of whatever you're referring to, and the URL where you found it). There's nothing dishonest about using information from any source, provided you give credit where credit is due.

This applies to the *Web*. Keep in mind, too, that the *Web* is increasingly used as a place to post initial drafts of research papers in order to get feedback from professional peers. This is a great way to find cutting-edge stuff: advanced drafts of papers by reputable scientists at reputable sites. But remember, these still haven't passed the quality control process, and they still need to be cited.

Mistake # 3: The *Web* has made the library obsolete. Yes, there's a tremendous amount of information out there, and someday it may be possible to do all of your research on the Internet. But, like those personal helicopters we're still waiting for, that's in the future. For now, the library is still the primary resource for learning and doing research, so don't get lulled into a false sense of security by the seemingly endless nature of the *Web*.

A related mistake is making a newsgroup your first stop in your search for information. If you follow a newsgroup very long, particularly those in the .sci category, you'll see at least one post by a student who has an assignment due the next day, desperately pleading for information on the topic at hand. If you follow the replies, you'll notice that, while they may be polite, they are not kind. Keep in mind that most people reading the list will be professionals who have devoted their lives to studying the topic. They tend to look unfavorably on lazy students' desperate requests for help, when it's clear that the poster hasn't bothered to look as far as a textbook or even an encyclopedia.

They may also be your instructor. A particularly funny (I thought) example came up a couple of years ago. A typically desperate student made a typically desperate post to a list. He received a prompt reply that went something like this: "For starters, come see me during my office hours and we can discuss how badly a flunking grade on this project will affect your grade in my course." I really felt for the poor guy, but he walked right into that one.

On the other hand, if you've done some research and have specific questions about something you don't understand, or there's no further information available on your campus, newsgroup queries are perfectly acceptable and you'll probably receive several helpful replies. Just keep in mind that your instructor is as likely to see your request as anybody else on the list, so conduct yourself accordingly.

Ok. End of sermon. Back to your browser!

Appendix I:
What's Under the Hood?

Here is a computer question to get your mind working. What is the difference between a Corvette and a Chevette? Okay, it's a stupid question but give it some consideration. The difference is a couple thousand dollars. Everything else is just fluff, mileage, and auto insurance. But, if we break this down to our basic needs (if we really need a ride) both cars are the same. The same goes for computers and networks. The simple no-gloss stuff will get you by while saving you money; and the high-gloss stuff will transform your cash into dash and make your Internet browsing a little more enjoyable. Many of you may be lucky enough to have computers on your campus that are set up to allow Internet access. In case you don't, I'll list the *minimum* systems, connections, and services that you'll need to enter the Internet, but you'll have to decide how shiny your system needs to be.

The Computer

Now watch how you approach this issue if you ask for advice on which computer platform to buy. Opinions are known to run a little hot with some people when they begin to argue the differences between Macintosh and PC-compatible systems. Proponents from both camps can easily get out of hand. If they begin to foam at the mouth, then you know you're in for a real adventure. The best advice that I can give to you is test them both at a computer store. Choose the one that you can pay for and are most comfortable using. After all, it won't do you any good if you don't enjoy using it or if it gets repossessed halfway through the semester.

These are the bare minimum configurations you'll need. That is, you'll be able to get by, but just barely, with these. As with most computer applications, you'll be happier with a more powerful processor and more RAM.

Macintosh

- 68020
- System 7.0
- 256 color monitor
- 8 MB of RAM
- 4 MB of free disk space for browser software

PC-Compatible

- Intel 386sx
- Windows 3.1
- VGA monitor
- 8 MB of RAM
- 4 MB of free disk space for browser software

The Internet Connection

If you get lucky, your campus has already recognized the importance of the Internet as a teaching and learning tool. If the equipment is set up, then they should have Internet, or more specifically *Web* access ready to go. Okay, if you didn't enroll in a campus with vision, bucks, or both, then you'll have to do it the hard way. There are many resources that can help you set up a connection from home.

Some campuses, although lacking a walk-in lab, have made arrangements for students to dial into the campus computer system and hopefully the Internet with a modem. If such is the case, then search out the campus computer hack and ask for help. This may require a fair deal of begging, bribing, or borrowing.

Another option for accessing the Internet is to subscribe to a company such as America Online, CompuServe, Delphi, eWorld, Prodigy, or Microsoft. Many other companies are also in business to sell you access to the Internet, and you should consider them in addition to these four. Some of these on-line service providers advertise the availability of access. However, you are required to use their software, and it may not have all of the options that you might want. As always, it is a buyer's market, and you should shop around and test drive everything before you buy.

When it comes to buying modems, I suggest you buy something as fast as possible but not a modem of less than 14,400 bits per second (bps). A bps is a measure of the speed with which the modem transfers data. The higher the number, usually the faster the transfer rate. With a typical 14.4K bps modem you can expect that it will take a few seconds to transfer a typical *Web* page, but this will vary depending on the complexity of the page. There are a couple of other things that come into play, but in general, a higher number means a faster modem.

The Browser Software

If everything has fallen into place, you should now have access to a computer with muscle and a gateway to the Internet, either through your school or an Internet service provider. But you're missing one very important piece, the *Web* client software. A more descriptive name for this software is *browser*, as that is what most people do with it. It is used to browse or wander, sometimes aimlessly, through the Internet.

There are many *Web* browsers on the market, and in the space of time that it's taken me to write this manual I would not doubt that several more have entered the race to capture your dollar. The mother of all browsers is NCSA's *Mosaic*, now in version 2.0, but there have been many others spawned from it. Although you might have contrasting opinions when you've tried a couple, I prefer Netscape Communication's *Netscape Navigator* (version 2.0 as I write this), and therefore, I've chosen to use it to illustrate this manual. All browsers have advantages and disadvantages. Although I prefer *Netscape* as my vehicle for Internet travel, you may find *Mosaic* or another to be more to your liking. I have included a list of many of the available browsers in Appendix II. You should really evaluate them all and choose the one that you're most comfortable with. Browsers are typically very cheap, if not simply free for educational use. (Considering the nature of my audience, I should give proper reverence to that last statement.) *Netscape*, *Mosaic,* and many of the other browsers are FREE for student use! So, don't be afraid to evaluate them all.

Now, if you have all of these basic elements and they've been put together correctly, you should be ready to surf. The rest of this manual is devoted to guiding you through some of the many strange and wonderful places that will allow you to enjoy the depth and uniqueness of the Earth sciences.

Appendix II:

Internet Resources

To access the URLs listed below, simply type the complete URL in the "Go To" box (usually at the top of the browser screen) and press "Enter." Remember: URLs are case sensitive and must be entered exactly as they appear below. Also keep in mind that URLs are likely to change, so you may need to do a search (use the title given below) to locate the resource. Your browser will probably come installed with its developer's home page as the default URL, and most of these (particularly *Netscape*) include links to extensive information about the WWW, including links to many of the resources listed below.

General Information on the *World Wide Web*.

Print Resources.

Keep in mind that, because the *Web* changes so rapidly, printed information about it (sadly, including this version of your manual!) is quickly outdated. The best sources of information are on the *Web* itself (these are also subject to obsolescence), but if you're more comfortable with a printed tome, try the following, or look for the most recent ones in your bookstore or library.

Kent, Peter. 1995. *The Complete Idiot's Guide to the World Wide Web.* Indianapolis: Alpha Books.

Stout, Rick. 1996. *The World Wide Web Complete Reference*. Berkeley: Osborne McGraw-Hill.

World Wide WebWeek (periodical). Westport, CT: Mecklermedia

Online Resources for Learning More About the WWW.

All of these sites contain information and links to other resources (including each other) explaining the *World Wide Web*, what's available on it, where to download software, etc.

WWW Frequently Asked Questions, maintained by Thomas Boutell (updated bi-weekly).
http://sunsite.unc.edu/boutell/faq/www_faq.html

Yahoo! World Wide Web information.
http://www.yahoo.com/Computers/World_Wide_Web/

NCSA: Introducing the *World Wide Web*.
http://www.ncsa.uiuc.edu/General/Internet/WWW/Intro/

University of Texas: Learning to *Web*
http://wwwhost.cc.utexas.edu/learn/

European Academic Research Network Association (EARN).
http://www.earn.net/gnrt/www.html

Hughes, Kevin. Entering the *World Wide Web*, A Guide to Cyberspace.
http://www.eit.com/web/www.guide/guide.toc.html or
http://www.hcc.hawaii.edu/guide.www.html

Internet tools.

An excellent source of information (brief descriptions and Net addresses for obtaining) on *Web* browsers, *spiders*, *FAQ*s and other information about the Internet is:

http://www.rpi.edu/Internet/Guides/decemj/itools/top.html

If you have access to FTP, the you can get the file at: ftp://ftp.rpi.edu/pub/communications/internet-tools.txt

Web Browsers.

A good list of browsers and information on them is available from *Yahoo!*:

http://www.yahoo.com/Computers/World_Wide_Web/Browsers/

You can obtain information for most of the browsers listed below, including instructions on downloading the software for evaluation, by sending an e-mail message to the addresses indicated (or by telephone).

Cello (Windows) by Cornell Law School Legal Information Institute. Internet browser with WWW, Gopher, FTP, Telnet, UseNet News. E-mail:
lii@fatty.law.cornell.edu

MacWeb (Macintosh) and WinWeb (for Windows) by EINet
800/844-4638; fax 512/338-3897
winWeb@einet.net

WinWeb:
http://www.einet.net/EINet/WinWeb/WinWebHome.html

MacWeb:
http://www.einet.net/EINet/MacWeb/MacWebHome.html

Internet-in-a-box (Windows) by Spry. Commercial product (can't be downloaded) available at retail outlets. Includes WWW, Gopher, FTP, Telnet, UseNet News.
info@ibox.com (type "help" in the body of the message)
http://www.spry.com/intabox.html

800-777-9638, ext. 26; fax 206/447-9008

Mosaic (many platforms) by National Center for Supercomputing Applications. The browser that started it all. A solid, if basic (by commercial standards) browser, with extensive documentation available via the WWW and FTP.
orders@ncsa.uiuc.edu

Home Page:
http://www.ncsa.uiuc.edu/SDG/Software/Mosaic/NCSAMosaicHome.html

Mosaic FAQ:

http://www.ncsa.uiuc.edu/SDG/Software/Mosaic/Docs/mosaic-faq.html

Downloading and running Mosaic

ftp://ftp.ncsa.uiuc.edu/Mosaic/README.Mosaic

217/244-3473

Netscape Navigator (MS Windows, Macintosh, and X) by Netscape Communications. Rapidly becoming the dominant *Web* browser on the market.

info@mcom.com
http://netscape.com/comprod/netscape_products.html

800/638-7483; fax 415/254-2601

SuperTCP by Frontier Technologies Corp. (Windows). Part of a suite of Internet tools, and the most full-featured browser available. Free evaluation; discount for educational site licenses.

superhighway@frontiertech.com
http://www.frontiertech.com
ftp://ftp.frontiertech.com

800-929-3054; Fax 414-241-7084

Web Publishing.

These sites have all the information you'll ever want on setting up a home page

Web Weavers

http://www.nas.nasa.gov/NAS/WebWeavers/

Homepage Construction Kit

http://nearnet.gnn.com/gnn/netizens/construction.html

How to Publish on the Net

http://www.thegiim.org/

Yale Center for Advanced Instructional Media Web Style Manual

http://info.med.yale.edu/caim/StyleManual_Top.HTML

Australia National University: Information Quality, Guidelines & Standards Home Page.

http://coombs.anu.edu.au/SpecialProj/QLTY/QltyHome.html

Software Archives.

The Virtual Software Library

http://vsl.cnet.com/

The Oakland Software Repository

http://www.acs.oakland.edu/oak.html

Appendix III:
A Call to Order

Internet Providers

America Online
America Online Inc.
8619 Westwood Center Dr.
Vienna, VA 22182-2285
800-827-6364
703-448-8700
aohotline@aol.com

CompuServe
CompuServe Inc.
5000 Arlington Centre Blvd.
Columbus, OH 43220
800-848-8199
614-457-8600
70006,101@compuserve.com

Delphi
Delphi Internet Services Corp.
1030 Massachusetts Ave.
Cambridge, MA 02138
800-695-4005
617-491-3342
617-491-6642 fax
askdelphi@delphi.com

Prodigy
Prodigy Services Co.
445 Hamilton Ave.
White Plains, NY 10601
800-776-3449

If you don't want to use one of these large service providers but would prefer a company that drops the frills and gives you direct access to the Internet, you can locate such service providers at this *Web* address (http://www2.celestin.com/pocia/domestic/d_rega.html). Just search by the area code for your phone service and you'll get a list of providers that will accept your money to use their service.

Appendix IV:

It's News to Me

Geography and Geology-Related Newsgroups and *Web* Sites.

Think of this as a starter kit to the Usenet. I've spent time browsing through many of the newsgroups on the Internet and put together the following list. This is not a complete list, and some of the newsgroups that I've listed will, without a doubt, disappear in the near future. Additionally, there will be many new groups born in this same time period. One of them might even be yours. As I've said before, it's a changing world.

The names are very descriptive for the topic of discussion. You will find that some are very interesting and some are not. Subscribe to several of the groups and eavesdrop on their conversations for a while before you jump in. This will give you some experience with the general attitude of the group and prevent some embarrassment on your part. I also suggest that you read the FAQ and any postings directed to new participants to the group.

alt.agriculture.misc
alt.culture*
alt.culture.theory
alt.folklore*
alt.great-lakes
alt.politics*
alt.society*
bit.listserv.geography
bit.listserv.history
bionet.agroforestry
comp.infosystems.gis
ncar.weather
info.grass.user
sci.agriculture
sci.anthropology*
sci.archaeology*
sci.bio.conservation
sci.bio.ecology
sci.bio.paleontology
sci.engnr.civil
sci.environment
sci.geo.earthquakes
sci.geo.eos
sci.geo.geology
sci.geo.meteorology
sci.geo.oceanography
sci.geo.rivers+lakes
sci.geo.satellite-van
sci.geo.image processing
sci.skeptic
soc.culture...**
soc.history*

* Also, newsgroups devoted to specific subcategories.
**140 subcategories, from ...afghanistan to ...zimbabwe

Web Sites

Here is a very small sample of interesting *Web* resources to get you started. Keep in mind that some of these sites may have changed URLs or may no longer be functioning. If the URL doesn't get you where you want to be, well, by now you should know what to do!

American Demographics Home Page
http://www.marketingtools.com

Centre for Landscape Research Network
http://www.clr.toronto.edu/clr.html

ESRI: Home of the GIS People
http://www.esri.com

History of Cartography Project
http://elvis.neep.wisc.edu/~cdean/index.html

The Human Language Page
http://www.willamette.edu/~tjones/Language-Page.html

US Bureau of Land Management (links to maps and GIS)
http://www.blm.gov/gis/gishome.html

The USGS home page:
http://www.usgs.gov/

The Virtual City
http://riceinfo.rice.edu/projects/RDA/VirtualCity/index.html

Two amazing places for finding all sorts of geology and geography Internet resources (including *Web* sites, mailing lists, software archives, etc. are:

The Virtual Earth:
http://atlas.es.mq.edu.au/users/pingram/v_earth.html

The Soft Earth:
http://atlas.es.mq.edu.au/users/pingram/s_earth.html

Appendix V:

Stepping Out

Student Homepage Template

```
<HTML>
<HEAD>
    <TITLE>Your Homepage</TITLE>
</HEAD>

<BODY>
<CENTER>
    <H1>Your Name</H1>
    <H2>Your Title, Major, or Philosophy</H2>
    <H3>
    <ADDRESS>
    Your address<BR>
    May<BR>
    Go<BR>
    Here<BR>
    <P>
    Your e-mail address<BR>
    </ADDRESS>
    </H3>
</CENTER>
<HR>

<DL>
    <H2>This Is Your Life:</H2>
    <DD>Just say something about yourself. After you've added some-
       thing to your homepage file, compare it to what actually
       shows on your browser. You'll notice that most of this gibber-
       ish is not visible.
    <P>
    <DD>You can have as many paragraphs as you wish. Here's
       another. Enjoy.
    <P>
</DL>
<HR>
<DL>
    <H2>Important Geoscience Resources</H2>
    <DD >
    <A HREF="http://atlas.es.mq.edu.au/users/pingram/v_earth.html">
    The Virtual Earth
```

```
    </A>
<DD>
<A HREF="http://atlas.es.mq.edu.au/users/pingram/s_earth.html">
    The Soft Earth
    </A>
<DD>
<A HREF="http://www.usgs.gov/">
    USGS Home Page
    </A>
<DD>
<A HREF="http://feature.geography.wisc.edu/phys.htm">
Physical Geography Resources
</A>from Univ. of Wisconsin Geography.
<DD>
    <A HREF="http://www.grdl.noaa.gov/">
    NOAA Geosciences Lab:
    </A>
<DD>
    <A HREF="http://www.ualberta.ca/~kbrett/index.html">
    Kevin's Page of Death
    </A>
<DD>
    <A HREF="http://www.utexas.edu/depts/grg/virtdept/resources/contents.html">
    U. Texas Virtual Department
    </A>
</DL>
<HR>
<DL>
    <H2>Misc. Internet Resources</H2>
    <DD>
        <A HREF="http://www.matisse.net/files/glossary.html">
        Glossary of Internet Terms
        </A>
    <DD>
        <A HREF="http://www.qualcomm.com/">
    Eudora Information
</A>
</DL>
<HR>
<DL>
    <H2>Directories and Search Engines</H2>
    <DD>
        <A HREF="http://harvest.cs.colorado.edu/">
        Harvest
        </A> (look for brokers)
    <DD>
        <A HREF="http://lycos.cs.cmu.edu">
        Lycos
        </A> home page. Great search engine.
    <DD>
        <A HREF="http://www.Webcrawler.com/">
        Web Crawler
        </A>
    <DD>
    <A HREF="http://www.cs.colorado.edu/home/mcbryan/WWWW.html">
        Web Worm
```

```
    </A>
    <DD>
        <A HREF="http://www.yahoo.com">
        Yahoo Directory
        </A>
</DL>
<HR>
<DL>
    <H2>Info on the Web and Homepage Design</H2>
    <DD>
        University of Texas <A HREF="http://wwwhost.cc.utexas.edu/learn/"> Learning to Web
        page.
        </A>
    <DD>
        <A HREF="http://www.ncsa.uiuc.edu/General/Internet/WWW/"> National
        Center for Supercomputing Applications</A> at the University of Illinois.
        </A>
    <DD>
        <A HREF="http://www.nas.nasa.gov/NAS/WebWeavers/">
        Web Weavers
        </A>
    <DD>
        <A HREF="http://nearnet.gnn.com/gnn/netizens/construction.html">
        Homepage Construction Kit
        </A>
    <DD>
        <A HREF="http://www.thegiim.org/ ">
        How to Publish on the Net
    </A>
    <DD>
        <A HREF="http://info.med.yale.edu/caim/StyleManual_Top.HTML">
        Yale Center for Advanced Instructional Media Web Style Manual
    </A>
    <DD>
        <A HREF="http://coombs.anu.edu.au/SpecialProj/QLTY/QltyHome.html">
        Australia Nat. Univ.: Information Quality, Guidelines & Standards
        </A>
    <DD>
    <A HREF="http://www.ncsa.uiuc.edu/General/Internet/WWW/HTMLPrimer.html">
    NCSA's Beginner's Guide to HTML</A>
</DL>
<HR>
<H3>
    If your homepage resides on a Web server, you can make it easy for people to e-mail you by
    clicking on: Your Name <A HREF="mailto:yourname@your_e-mail_address">E-mail address</A>.
</H3>

</BODY>
<HTML>
```

Glossary

It's All Greek to Me

Bookmark

A list of URLs saved in a browser. By saving URLs in a bookmark list, you can readily access the information resource it refers to by pointing and clicking on the highlighted bookmark. Some browsers (e.g. *Mosaic*) use the term "Hotlist."

Browser

Software program used to navigate and view information on the Internet; a particular type of "Client."

Browsing

Navigating the Internet using Gopher or a *Web* browser to jump directly from one information resource to another, as opposed to connecting with and logging onto each computer independently.

Cache

Temporarily stores information on a *Web* page in either RAM or your hard disk. The browser can access and display cached pages much more rapidly than if it has to retrieve them from the server each time you wish to view them.

Clickable image (clickable map, mapped image)

An interface used in *Web* documents that allows the user to click on part of an image and get a response.

Client

Software program for accessing information from a remote computer. Also used to refer to the computer requesting information from a server in a *client-server* information exchange model.

Client-server model

A model of information exchange in which most a *client* computer requests and receives information from a *server*, where most of the back-end processing (e.g. database searches) is performed.

Compressed file

File that has been written to a storage device in a format that minimizes its size (in bytes). Common formats include *ZIP* and *LZS*. If you download a compressed file, you must expand it using a file compression utility such as *PKUNZIP* (for *ZIP* files).

Discussion Group

(also "list," "listserv," "newsgroup," or sometimes "mailing list"). A group of people with similar interests who post and read messages to a list server dedicated to a particular topic. Some lists are moderated (i.e., posted messages are reviewed by a moderator to ensure that they are appropriate to the list) before being passed on to subscribers. Others are not.

Download

Process of transferring a computer file from a remote computer. (See UPLOAD.)

Extensions

Codes appended to file names that indicate the file format. Common file extensions include:

.aiff	AIFF sound
.au	AU sound
.gif	GIF image
.html	HTML document
.jpg or .jpeg	JPEG image
.mov	QuickTime movie
.mpeg or .mpg	MPEG movie
.ps	PostScript file
.rft	Richly Formatted Text
.tiff	TIFF image
.txt	Plain text
.xbm	XBM bitmap image

File compression saves storage space and speeds the transfer of files. File extensions indicate what compression utility was used to archive the file, and thus what you need to uncompress it. Compression formats and programs to decompress them (in italics) include:

.arc	An older DOS format (*ARC*)
.exe	self extracting archive. Simply execute the program to decompress.
.gz	variation of .Z (*GNUZIP*)
.hqx	binary to 8 bit ascii encoded. (*Binhex*)
.lzh	An older DOS format (*LHARC*)
.sea	self extracting archive (MacIntosh)
.sit	Stuffit compression (*Stuffit* for Mac; DOS & UNIX versions available)
.tar	tape archive (Unix *tar* utility)
.Z	Unix compression (Unix *unpack* utility: DOS & Mac unpack and unzip utilities available)
.zip	ZIP compression (*PKUNZIP*)

FAQ

Frequently **A**sked **Q**uestions, a document posted by network administrators or moderators containing answers to common questions about a subject.

Forms

An interface element that allows users to enter information into certain parts of a *Web* page and submit it to the server for processing.

FTP

This stands for File Transfer Protocol. It is a procedure used to transfer large files and programs from one computer to another. Access to the computer to transfer files may or may not require a password. Some FTP servers are set up to allow public access by anonymous log-on. This process is referred to as *Anonymous FTP*.

GIF

(**G**raphics **I**nterchange **F**ormat). A graphic file format created by CompuServe for transferring digital images electronically. GIF files can be identified by the file extension .GIF appended to the file name.

Gophe.

A format structure and software tool for Internet communication, navigation, and file retrieval.

Graphics file

File containing a digitized image. Graphics files on the Internet are most commonly in GIF or JPEG format. In order to view graphics files, you must have a viewer installed on your computer.

GUI

Graphical **U**ser **I**nterface. Refers both to the appearance and method of interacting with a computer. The most common tool for interacting with a GUI is a mouse, which is used to select icons representing commands. Macintosh, Microsoft Windows, and X Windows are examples of Graphical User Interfaces.

Helper application

An auxiliary program that allows a *Web* browser to read and use file formats that it cannot handle internally.

Home page

The *Web* document consisting of text, images, and links that a browser loads and displays as its primary point for navigating the Internet. More generally, the term refers to a *Web* site and is sometimes used to refer to any *Web* document.

Host name

An alias for an IP address consisting of a machine name, a domain name, and a domain identifier.

Hotlist

See BROWSER.

HTML

Hyper**T**ext **M**arkup **L**anguage, the simple scripting language used to write *Web* documents. Some browsers support expanded versions of HTML that support more elaborate or sophisticated *Web* page design.

HTTP

Hyper**T**ext **T**ransport **P**rotocol. The protocol governing communications between *Web* servers.

Hypertext

Nonsequential text retrieval. Hypertext documents, such as many *Web* documents, have embedded links (or "hyperlinks") to other information resources on the Internet.

Information resource

Essentially, anything that can be accessed on the Internet. This includes text, data, image, audio, and video files, but the term refers more generally to networks and servers accessible via the Internet as well.

Inline Image

Images viewed along with text in *Web* documents. Inline images are stored in GIF format, and do not require use of an external viewer (see *helper applications*).

IP

(**I**nternet **P**rotocol) **address**. A 32 bit numerical address consisting of four 8 bit numbers (0 to 255) separated by periods, called "dots" in Net jargon (e.g. 123.76.112.23) that identifies a particular internet host.

JPEG

(**J**oint **P**hotographic **E**xperts **G**roup). Also refers to a graphics file format for transferring digital images.

List server

An Internet computer on which the e-mail addresses and often archived messages related to a *discussion group* are stored, and that handles the receipt and distribution of messages posted to the group.

MIME

(**M**ultipurpose **I**nternet **M**ail **E**xtensions). A standardized system for organizing different file formats by MIME type (e.g. compressed or graphic file formats).

Mosaic

First *Web* browser with a consistent interface across Macintosh, Unix, and Windows environments, developed by the National Center for Supercomputing Applications and subsequently licensed to a number of commercial firms that have created enhanced versions.

MPEG

(**M**otion **P**icture **E**xperts **G**roup). Also refers to a file format for playing digital audio and video on the computer.

Newsgroup

Discussion groups, especially on the Usenet News, but more generally any *discussion group*.

Protocol
A common network language consisting of a list of parameters and standards governing information transfer among computers.

QuickTime
A format for playing audio and video files, used by Apple Computer.

Server
The computer that processes and provides information from a client in a client-server information exchange model. A software program used to provide information to remote computers.

Site
1) A computer or network connected to the Internet with a server configured to reply to requests clients. 2) A WWW site is a collection of information resources linked together by a common home page with a URL that makes it accessible via the Internet.

Spider
A software program that traverses the WWW, visits network hosts and gathers information about resources they contain. Also called crawlers, robots, worms, etc.

Subscriber
Somebody whose e-mail address is listed for distribution in a discussion group.

Upload
Transferring a computer file from one computer (usually a client) to another (a server).

URL
Uniform **R**esource **L**ocator. An internet address using a common format that describes: 1) the protocol for transferring information; 2) the host computer address; 3) the path to the desired resource; and 4) the name of the file (optional). (See page **Error! Bookmark not defined.)**

Usenet
An international network for distributing news and communications within discussion groups.

Viewer
Software that displays digital images on the computer screen.

WAIS
(**W**ide **A**rea **I**nformation **S**ervers). A system that allows users to search indexes of information on the Internet.

WAV
Wavform sound format, a Microsoft Corporation format for encoding and transferring sound files.

Suggested Reading

For more information on surfing the *Web* and creating your own *Web* page, check out these exciting Internet titles:

Web Surfing

Clark, Michael, *Cultural Treasures of the Internet*
The Internet guide for everyone interested in the Arts and Humanities.

Comer, Douglas, *The Internet Book*
A unique overview of the Internet and networking technology and terminology.

Fristrup, Jenny, *The Essential Web Surfer Survival Guide*
A detailed resource guide to "reach" the *Web* quickly and easily.

Kehoe, Brendan, *Zen and the Art of the Internet*
The legendary newcomer's guide is now fully updated!

Web Programming

Morris, Mary *HTML For Fun and Profit*
Bestseller now includes *Netscape* extensions and Windows server!

Neou, Vivian, *HTML CD for Windows*
All the software you need to transform your PC into a *Web* publishing system.

Neou,Vivian, *HTML CD 3.0 with JavaScript for the Mac*
The most complete *Web* publishing toolkit for the Mac available!

Pew, John *Instant Java*
The quickest and easiest way to add the power of JAVA to your *Web* pages.

To order any of these titles, call the Prentice Hall Order Processing Center: 1(800) 947-7700 or visit your local Prentice Hall PTR Magnet Store. For the store nearest to you visit us at:

http://www.prenhall.com/